What is Aromatherapy?

Aromatherapy can reduce stress, improve sleep, and give you more energy. Aromatherapy can change moods, alter perceptions, and improve digestion. It can improve your complexion, or eliminate a sinus headache. Aromatherapy is the use of fragrance to improve health.

Merely smelling a fragrance influences us physically and mentally, alters hormone production and brain chemistry, changes metabolism and affects our thoughts and emotions.

Few things can impact us so deeply as specific smells. Smells can take us back to our childhood, conjure up a lost love, or evoke old emotions as real as the day we experienced them. Smell is the most direct of all of our senses and has an immediate impact uninfluenced by language, culture, or the passage of time. We smell with every breath we take and we are continuously monitoring the world around us subconsciously.

Incorporating aromatherapy into your daily routine can provide numerous natural health benefits such as improving mood, decreasing stress and anxiety, improving appetite and digestion, alleviating fatigue, and improving sleep.

There are two main ways to use fragrance therapeutically. One is through inhalation alone, which has significant impact on mood and emotions but also produces physical reactions. The second route is through physical application of essential oils to the body, such as through a massage or bath.

Essential Oils

Squeeze a garlic clove or some basil leaves and then smell your fingers. That aroma is the result of volatile molecules released by bursting tiny glands in the plant material.

All plants produce at least two types of oils. The oils we use for cooking like olive oil or sesame oil are considered base oils. These stable oils are not very aromatic and do not evaporate quickly. These oils are often produced by the plant for reproduction and are usually found in the seeds.

Plants also produce another oil, naturally fragrant, and this is the oil that we use in Aromatherapy. These oils are produced in various parts of all plants. They are found in fruit peels (all citrus), tree resins (frankincense and myrrh), flowers (rose and lavender), leaves (sage and peppermint), bark (cinnamon and sassafras), grasses (lemongrass), rhizomes (ginger), and seeds (fennel, dill). Many plants make many different types of essential oils in different parts of the plant such as bitter orange, which produces petitgrain in the leaves, neroli from the blossom, and bitter orange from the fruit peel.

To early herbalists and healers these oils represented the spirit of the plant, or the essence. Hence these oils are called essential oils.

Plants produce these oils for different reasons. Some of them transmit nutrients across cell walls throughout the plant, some protect against predators or act as antibacterial agents, some attract insects for pollination.

Essential oils are produced by specialized plant structures called secretory cells that trap the photoelectromagnetic energy of the sun and combine it with glucose to produce biochemical energy in the form of aroma molecules. These molecules are extremely volatile and will rapidly fly off in every direction as soon as they are released from the glands they are contained in within the plant, which is why we are able to smell them and why they so rapidly evaporate in an open container.

There are about 30,000 known aromatic molecules. The number and combination of these molecules is tremendously different from plant to plant making each essential oil completely unique. Any given drop of pure essential oil has a complex array of components (alcohols, phenols, esters, ketones, aldehydes, oxides, etc.) that gives each plant its character, odor, and flavor, and accounts for that particular oils' unique effects on our bodies.

Essential oils are extremely concentrated. Most of them are more than 50 times more potent than the plant from which they are derived. In most cases each drop of essential oil represents the potency of more than one ounce of plant material, and sometimes several ounces of delicate flowers are required to make just one drop of oil.

Ever since it was discovered that fragrance could be isolated from plants, human beings have experimented with the best way to do this.

Distillation

About 80 percent of the oils produced are extracted through distillation. For hundreds of years it has been the best method of extracting pure oils and it still produces a very pure product.

During distillation plants are exposed to boiling water or rising steam and release their fragrant oils into the steam. The oil laden steam is moved to another area and cooled so that the oil and water returns to liquid form. Because water and oil do not mix the essential oils can be easily separated out.

The amount of oil produced depends upon the length of distillation, the temperature, and the type and quality of the plant material. On average it takes well over 500 pounds of plant material to produce 32 ounces of essential oil.

Cold Expression

During cold expression, or scarification, the plant is shredded and then mechanically pressed under tremendous pressure. The resulting emulsion is place into a centrifuge and the essential oil is separated out.

Solvent Extraction

Solvent extraction made it possible to create a whole range of oils that could not be produced before because their structures were too delicate to withstand the other methods of extraction. Solvents do not require heat, or pressure, or water for extraction. However, these methods are not without drawbacks. The solvents used for extraction (butane, hexane, acetone, benzene, propane, methane) are toxic and some are even carcinogenic.

The plant material is submerged and agitated with the solvent which dissolves the volatile oils as well as some waxes and pigments. The solvent is then evaporated off under pressure. The result is a soft sticky wax called a *concrete*. The concrete is then mixed with ethyl alcohol and chilled. The alcohol is then removed and the resulting product is called an *absolute*.

Absolutes are said to be relatively free of the solvents and alcohol used during extraction but some oils contain slight residues in the final product. New methods are being developed which hold the promise of being able to remove all solvents.

Carbon-Dioxide Extraction

This process allows for high pressure low heat extractions to be completed on even the most delicate and volatile oils in just a few minute completely solvent free. Because the solvent is a gas it is easily and totally removed when the pressure is released.

CO_2 extracts provide an exact representation of the plants smell and taste, which even the best distillation methods cannot match. Until recently these extracts were rarely available to the consumer, and even now these products are much more expensive than their distilled counterparts.

Quality and Purity

Quality and Purity are often regard as the same thing, but they are not.

Quality relates to the "grade" of an oil, which is influenced by growing conditions, nutrient levels, and processing or extraction methods.

Purity refers to authenticity, that the product is unadulterated.

Because most essential oils are grown outside the United States it is sometimes difficult to confirm both quality and purity, and the costs can be quite high due to the rapidly increasing demand. This is why so many cosmetic manufacturers use synthetics. Most fragrances used in cosmetics and skin care products are synthetic, which cuts costs and provides a consistent and controllable aroma.

Synthetics

If the label on a product says "fragrance" or "perfume" it is most likely synthetic. Synthetic aromas are chemically manufactured from a variety of easily manipulated base oils like petroleum or pine oil.

Synthetic fragrances are not suitable for Aromatherapy because they simply mimic a scent but do not contain the chemical constituents that give essential oils their characteristics. The potent aromatic chemicals used in synthetics must be extended in a solvent base in order to replicate the essential oil as closely as possible. Both the solvent and the chemicals used may be toxic and can be absorbed through the skin.

How to tell the Difference

Many oils cannot be produced naturally (Hibiscus, Apple, Lily-of-the-Valley, Lotus) or can only be produced through very expensive CO_2 extraction techniques, so if you see them they are synthetics. If you find a selection of these oils, be suspicious of all of the other oils from that brand.

Use the price as a yardstick. Natural extracts are very expensive, and with high quality plant material and better extraction methods the cost goes up. Be wary of oils priced significantly lower than comparable products.

Because you cannot always trust a label or a store clerk, the best indicator of quality is ultimately your nose. Becoming a good judge of quality isn't that difficult and even inexperienced people can identify synthetics right away when given scents to compare. Therefore, training your nose is probably your best means of monitoring quality.

The History of Fragrance

Ancient perfumery began with the burning of gums and resins. Many cultures soon began incorporating richly scented plants into animal fats to anoint the body for ceremony and pleasure. Humanities earliest written documents record the use of plant oils for healing. Oils, ointments, infusions, poultices and incense were the only medicines known until the end of the 19th century.

4000 BC

As early as 4000 BC people throughout the African continent coated their skin with fragrant oils to protect themselves from the hot, dry sun. Resins and gums were combined with animal fat and honey to produce incense.

3000 BC

In 3000 BC when the Egyptians were learning to make bricks, they were already importing enormous quantities of spices, gums, and fragrant plants, primarily myrrh. Most of these were used in religious ceremonies and incense was burned at all important events. Papyrus manuscripts recorded the use of aromatics for magical, mystical, and healing experiences. Incense was involved in exorcisms, healing, and in ritual lovemaking for weddings. It was believed that incense smoke could attract and please the gods, and bring good luck.

In China the Taoists believed that extracting a plants' fragrance represented the liberation of the soul. The Chinese Yellow Emperor Book of Internal medicine was published in 2697 BC and details the use of aromatic plants for medicine and for controlling moods.

2000 BC

The Vedic texts of India, which date around 2000 BC list numerous plant oils as healing and spiritual substances. Aromatics were considered spiritual and handling herbs and oils was a sacred task. Among those listed were cinnamon, ginger, spikenard, sandalwood, myrrh, jasmine, patchouli, and rose.

The Indians also made love potions of amber and wine that contained the essential oils of rose, sandalwood, frangipani, tuberose, champa, and mogra.

A still used for the distillation of essential oils was found in Afghanistan that dates back to 2000 BC, and tablets found in modern Iraq describe Mesopotamian ceremonies that use incense of cedar and myrrh to please the gods.

Around 1700 BC the trade routes expanded and the principle items of trade were spices, incense, and oils. These trade routes throughout the Middle East would be used for over 30 centuries, until the Portuguese discovered a way around Africa.

The Ancient Hebrews used fragrance to consecrate their temples and altars and for anointing and initiating priests. The Old Testament of the Bible lists the uses of many aromatics and gives recipes for making holy anointing oils. When the Jews returned from slavery under the Babylonians they brought back a heightened use of incense and distant fragrances.

1000 BC

The Greek world was rich in fragrance. At Delphi oracle priestesses sat over smoldering bay leaves to inspire intoxicating trances. Hundreds of perfumers had set up shop in Athens by the first century BC, and sold expensive oils in elaborately decorated clay or ceramic pots. Trade was heavy in marjoram, lily, thyme, sage, anise, rose, and iris. These fragrances were often infused into the oils of olive, castor, or linseed to make thick unguents. Mediterranean athletes often competed nude with these thickly scented unguents and waxes upon their head.

100 AD

By the First century AD Rome was going through 2800 tons of Frankincense and 500 tons of myrrh each year. The Roman Emperor Nero had the ceilings of his palace fitted with concealed pipes that sprayed down fragrant waters on the guests below, while panels slid aside to shower people with clouds of rose petals.

Both men and women literally bathed in perfume, while attended by slaves called the *cosmetae* (where we get our word cosmetics from).   The first written descriptions of a still come from this time. Initially stills were used to distill essential oils, but later proved useful to alcoholic beverages.

Fragrance occurs throughout the New Testament as well, the gifts of Frankincense and myrrh given to the Christ child were more valuable than the gift of gold. Jesus anointed Mary Magdalene's feet with costly spikenard, and the word Christ comes from the Greek *Christos* which means anointed one. Gnostic Christians were deeply influenced by Egyptian philosophy and held fragrance in high regard and embraced the symbology of essential oils.

Socrates heartily disapproved of perfume, worrying that it might blend the distinctions between slaves (who smelled of sweat) and free men (who apparently did not). Alexander the Great developed a love for fragrance after his travels in Asia. He expanded trade with Yemen and Oman and constantly burned Arabian incense next to his throne. He established a botanical gardens in Athens and brought back fragrant plants from all of his travels.

The Japanese had already perfected the still at this point, but incense did not arrive in Japan until the 1rst century AD. Special schools were set up to teach *Kodo*, the art of perfumery. Powdered herbs were pressed into pastes of seaweed, charcoal, plum pulp, and salt, and then burned ceremoniously on beds of ashes. Elaborate story dances were told during incense ceremonies, and several schools still teach kodo today.

The Aztecs were as extravagant with fragrance as the Egyptians, and they too manufactured ornate vessel in which to burn incense. Injured Aztec were massaged with scented salves and placed in sweat lodges called *temazcalli*. The Mayans had similar clay sweat lodges, and the Incas made massage ointments of valerian and seaweed. Fragrant perfumes made from Balsam, juniper, sassafras, and vanilla were found throughout all of these cultures.

North American natives used smoke from burning plants or steam made by throwing herb-infused waters over hot rocks to treat many illnesses. Sick people were smudged with tight bundles of fragrant herbs and sweet grass, and Echinacea was smoked for headaches.

The expansion of Islam into parts of Europe and Arabia also spread the knowledge of fragrance. Mohammed's favorite scent was henna but it was the rose that came to permeate Muslim culture. Rosewater purified mosques, scented gloves, flavored sherbet and Turkish delight, and was sprinkled on guests. Prayer beads made from rose became popular at this time.

The Chinese upper classes made lavish use of fragrance during the T'ang Dynasties which began in the 7th century and continued to do so until the end of the Ming dynasty in the 17th century. Their baths and bodies were richly scented, as were their homes and temples. Their garments were fitted with little pockets which were filled with sachets of fragrant herbs and flowers, and even the ink was scented before being used on the scented paper. Their cosmetics and makeups were fragrant and the ribs of the fans they carried were made from fragrant sandalwood. Spectators at dances and ceremonies were showered with perfumed sachets and statues of Buddha were often carved of fragrant camphor wood.

Throughout the dark ages in Europe, nuns and monks kept secret formulas for making medicines from herbs and oils. In the 9th century the Medical School of Salernum Italy opened, and it was here that much of the knowledge that was preserved by the Muslims after the fall of Alexandria was reestablished in the west. This school's texts were used extensively for over 6 centuries.

By the 13th century Italy monopolized the Eastern trade routes and controlled the enormous importation of spices used to disinfect cities against the plague. The Italians were forced to use Muslim middle men who charged a 300% mark-up, so Marco Polo traveled to China to try to convince the Orient to trade directly with Italy.

Christopher Columbus's trip to the to the New World was an attempt by Spain to cut Italy off and find a shorter route to China. His discovery of tobacco, cocoa, vanilla, and chilies immediately started a trade of these goods.

But it was the Portuguese who finally discovered a route around the tip of Africa and gained control of the spice trade. Italy persuaded the Muslims to fight the Portuguese but their attempts were unsuccessful. The number of distilled and medicinal plants available in Europe during this time expanded dramatically.

People sniffed pomander balls and boxes full of cedarwood to ward off diseases. Many perfumers and glove makers escaped the plague due to the antibiotic and antiviral properties of the oils they were exposed to daily.

In the 18th century England finally got into the action by exploiting the tensions between the Muslims and the Hindus and converting much of India into a British colony. The British published extensive volumes of the medicinal and fragrant botanicals of India.

The Modern World

But at the 1867 Paris International Exhibition something happened which changed the use of fragrance dramatically. For the first time ever, perfumes were exhibited as separate from medicines. The first synthetic fragrance (coumarin) was also introduced which marked the first perfumes unsuitable for medicinal use. The cosmetics industry was born and the concept spread like wildfire, and for 50 years the use of fragrance in medicine was forgotten.

In 1928 French chemist Rene-Maurice Gattefosse coined the term Aromatherapy. He had severely burned his hand and plunged it into the only thing near, a container of lavender oil, and was amazed at how quickly his hand had healed. In World War II Madame Marguerite Maury and Dr. Jean Valnet experimented with herbs and essential oils to treat wounds and burns and found fragrance very effective at treating psychiatric problems. His book, the practice of Aromatherapy created a modern movement in Europe. That book greatly influenced an American Massage therapist Robert Tisserand to write his own book, the Art of Aromatherapy that sparked a similar interest in America in 1977.

Modern medicine is now rediscovering the uses of fragrance from ancient times, and a revival of Aromatherapy is sweeping the across the world.

How does fragrance affect us?

To fully understand how essential oils work we must have a basic understanding of two physiological processes, how our body processes odors and how oils are absorbed into the body.

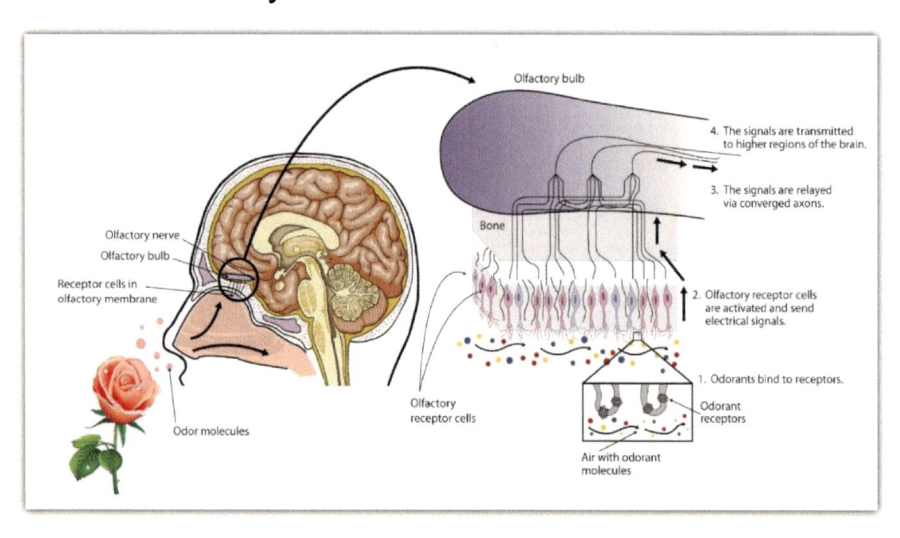

Odors are the effect of volatile molecules flying about in the air and entering our nostrils as we inhale. Then our body goes through 3 steps.

Reception takes place when an odor binds to a receptor cell with about 20 million nerve endings.

Transmission occurs when those receptor nerves transmit a signal to the olfactory bulbs at the base of the brain. There the signal is amplified and interpreted, and a message is sent to the limbic system.

Perception takes place when the hypothalamus receives the information and signals the pituitary gland (to send chemical messages to the bloodstream), the olfactory cortex (to identify the odor), the thalamus (to connect the odor to thought functions), the neocortex (to relate that scent information to information from the other senses), and to the higher brain function that stimulate conscious thought. These 3 steps happen in just a fraction of a second.

Smell is our most direct means of communication with the outside world.

We smell with every breath we take, and without even being aware we are doing so. Just eight molecules of a fragrance can create changes in our brain and activate nerve endings, but it takes a minimum of 40 nerve endings to fire at once before we become consciously aware of a fragrance.

Our eyes are able to distinguish several thousand colors with only 3 types of receptors, and yet we have over 1000 types of receptors for odor! Our brain can process over 10,000 different odors in an area of the brain about one inch square.

Our sense of smell is vital to our ability to taste. The tongue perceives only sweet, sour, and salty. All other tastes are actually odors. Studies done at the University of Texas identified as many as 20 different taste sensations that are actually odors, including bitter, some types of sweet, cool, metallic, and tangy.

Because smell is our only sense with nerve endings in direct contact with the outside world, odors can directly affect the mind. The brain is protected by a layer of fatty oils called the lipid layer, or blood-brain barrier. Oxygen and some nutrients can pass through this membrane, but larger molecules (like those of most medical drugs) cannot. This makes treating mental diseases with medications very difficult. But the olfactory nerves evolved before the brain, and they are not protected by this sheath. Odors can therefore directly affect the mind and the nervous system, without any filtration.

Our brains response to odors is influenced by our thought patterns and emotions, and by previous experiences. Certain brainwaves are very sensitive to emotional changes and are activated by specific types of fragrances. Different odors can stimulate different areas of the brain which in turn releases different types of neurotransmitters which then can affect every function of the body.

If the thalamus is stimulated (clary sage or grapefruit) neurochemicals called enkalphins are released which are natural pain-killers and cause feelings of well-being. Odors that stimulate the pituitary gland (jasmine, ylang-ylang) can regulate other glands, like the thyroid and the adrenals, and act as an aphrodisiac. Odors (like marjoram) that activate the "raphe nucleus" trigger the secretion of serotonin, which can promote sleep.

Essential oils are designed by plants to carry nutrients across cell membranes, and because of their extremely small size and lipid solubility they easily pass directly through our skin as well. Once through the skin essential oils enter the blood stream and pass rapidly throughout the body.

You can easily test this at home. Take a fresh clove of garlic and crush it. Rub the crushed garlic on the bottom of your bare foot, and within 20 minutes you can taste it. Essential oils administered to the skin can be smelled on the breath within minutes and all traces are eliminated from the blood in less than 2 hours.

Studies done after aromatherapy massages showed that 20 minutes after application to the body concentrations of the essential oils were highest in the blood, and then those oils diminished to indistinguishable levels after 90 minutes.

During the Black Plague, stench was linked to disease and smell was used in the diagnosis of disease. Typhoid fever smells like freshly baked bread, diabetes like sugar, the plague like apples, nephritis like ammonia, and yellow fever like a butcher shop. We now have computer profiles to help use odors to identify infections. Specific odors provide pathways through the central nervous system to activate immune protector cells, and researchers believe it may be possible to teach the body to activate its immune system by sniffing certain odors.

Anosmia is the term for loss of smell. It is estimated that over 2 million Americans suffer from an inability to smell or taste. Olfactory deficits can be caused by smoking, exposure to toxic chemicals, mineral deficiencies, hormones, radiation, diet, injury, or viral infections. Supplements of zinc may be helpful in restoring the sense of smell if deficiency is the cause.

The Biology of Love and Odor

The word kiss originally represented a desire to linger where anothers' smell was, and in some languages the word means a prolonged sniffing.

Each of us has an individual personal odor that is as unique as your fingerprints. Our personal odor is continually changing with or diet, health occupation, emotional state, and mood. It is influenced by our gender, our age, our heredity, and our sexuality. Personal odor communicates who we are and instinctively, unconsciously, we use this information as part of the criteria for choosing friends and lovers.

People who don't like each other's smell may have some level of incompatibility (physical or psychological) and will probably not make it as a couple.

Napoleon once sent a message to Josephine that read "Home in 3 days, don't wash." Goethe carried around the unwashed bodice of his lover so as to be near to her fragrance. In Elizabethan times lovers exchanged peeled apples drenched in sweat called "love apples" so that they could inhale the fragrance when they were apart. In Austria young men danced with a handkerchief in their armpit and then presented it to the girl he admired.

Odiferous substances manufactured by the apocrine glands found on your ears, eyelids, face, scalp, armpits, nipples, anus, and genital regions are called pheromones. This is from the Greek words *pherein* (to carry) and *hormon* (to excite). These glands become active at puberty and play a vital role in sexual behavior, puberty, menstruation, and menopause. Before puberty, perspiration has no odor as there is no need to determine sexual compatibility or find a mate.

Studies show that girls who are separated from boys during adolescence with generally start menstruating later than girls in co-ed environments. This is because contact with pheromones from boys going through puberty can trigger girls to become fertile. Women produce pheromones that will cause nearby women to synchronize with their menstrual cycle after 3 or 4 months. The male underarm scent can also regulate a woman' cycle if there is daily prolonged contact. Women who regularly have sex with the same partner will generally have cycles of shorter length, lesser intensity, have fewer infertility problems, and have milder menopauses than women who have infrequent or irregular sex, sex with different partners, or are celibate. All of these phenomena are due to odor and the sense of smell. Scientists are isolating the smells responsible for these phenomena in order to create scent-based birth control and cycle regulation.

After menopause or hysterectomy women lose their ability to detect musk odors, which smell similar to human testosterone. 25% of women with smell disorders lose interest in sex. Women's sense of smell is much stronger than a man's, and sense of smell increases during ovulation. During ovulation a woman's sense of smell can be up to 10, 000 times that of a man's. Even as infants, females have in general a much keener sense of smell then males.

      Smells can affect our judgement as well. Men perform better at mental tasks in unscented rooms, women perform better scented rooms. Male interviewers rated scented female applicants as more attractive and less intelligent, but when interviewed by women the scented applicants were seen as more intelligent and friendlier.

Guidelines for using Essential Oils

Essential oils are very versatile and can be used in many ways, but because these oils are very concentrated and highly potent, some knowledge of how to use them safely is vital to your success. The potential hazards of an oil depend upon the compounds on the oil, the dosage and frequency of use, and the method of application. Here are a few basic guidelines for the safe and effective use of essential oils.

Don't use undiluted essential oils on your skin

Use only pure essential oils from plants

Test for sensitivities

Do not use photosensitizing oils on the skin

Keep all essential oils away from eyes

Keep all essential oils away from children

Change the oils you use frequently

Don't take undiluted essential oils orally

Avoid essential oils during the first trimester of pregnancy, and use only nontoxic oils for the remainder of pregnancy

Avoid overexposure

Photosensitizing Essential Oils	
angelica	lime
bergamot	lemon
bitter orange	rue
cumin	verbena

Mucous-Membrane Irritants	
allspice	savory
cinnamon	spearmint
clove	thyme
oregano	

Skin Irritants	
cinnamon	Pimento
clove	savory
tea tree	thyme
oregano	Wintergreen

Toxic Essential Oils

Some of the oils in the following list have limited use externally, in small areas, and for short durations. The Latin names have been included to avoid any confusion. I recommend not using these oils at all until you have the proper training.

Mildly toxic essential oils

almond, bitter (Prunus amygdalus var. amara)
inula (Inula graveolens)
khella (Ammi visnaga)
mugwort (Artemesia vulgaris)
pennyroyal (Mentha pelugium)
sassafras (Sassafras albidum)
thuja (Thuja occidentalis)
wintergreen (Gaultheria procumbens)

Very toxic essential oils

ajowan (Ptychotis ajowan, Carum ajowan)
arnica (Arnica montana)
boldo (Peumus boldus)
buchu (Barosma betulina)
calamus (Acorus calamus)
cascarilla (Croton eluteria)
chervil (Anthriscus cerefolium)
camphor, brown and yellow (Cinnamomun camphora)
deer tongue (Carphephorus odoratissimus)
horseradish (Cochlearia armoracia, Armoracia rusticana)
jaborandi (Pilocarpus jaborandi)
mustard (Brassica nigra)
narcissus (Narcissus poeticus)
nutmeg (Myristica fragrans)
parsley (Petroselinum sativum, Carum sativum)
rue (Ruta graveolens)
santolina (Santolina chamaecyparissus)
Spanish broom (Spartium junceum)
tansy (Tanacetum vulgare)
tonka (Dipteryx odorata)
turmeric (Curcuma longa)
wormseed (Chenopodium ambrosioides, C. anthelminticum)
wormwood (Artemisia absinthium)

Carrier Oils

Carrier oils are made of larger molecules that do not easily pass through the skin, are stable and liquid at room temperature, and are much less aromatic than essential oils. These fatty oils are high in vitamins A, E, and F, and nourish and soften the skin, which makes them ideal carriers for essential oils. Different levels of saturated fats from plant to plant provide a wide variety of carrier oils that have different thicknesses, smell, and color. The more saturated an oil is the thicker its consistency and the longer it can be stored without refrigeration.

These carrier oils are often called fixed oils because the large molecules that make them up are not easily released from the plant. Because of this reason many of these oils are heat extracted or extracted with solvents. Whenever possible try to use oils the have been cold-pressed or expeller pressed, because solvents and heat can leave toxic residues in the oil.

Experiment with carrier oils of different consistencies to change the thickness, smell, and uses of your aromatherapy products.

Methods of Application

Dilutions

The most effective way to dilute essential oils is in a carrier oil. A carrier can be any high-quality vegetable oil, such as almond, apricot, hazelnut, olive, grapeseed or sesame. Carrier oils are stable and liquid at room temperature, do not evaporate as quickly and are not nearly as aromatic.

A safe and effective dilution for most aromatherapy applications is 2 percent, which translates to 2 drops of essential oil per 100 drops of carrier oil. There is no need to go beyond a 3-percent dilution for any purpose. In aromatherapy, more is not better; in fact, "more" may cause adverse reactions. Some oils, such as lavender, are sedating in low dilutions and stimulating in high dilutions.

A 1-percent dilution should be used on children, pregnant women, the elderly, and those with health concerns.

1% dilution:	5 drops essential oil per ounce of carrier oil
2% dilution:	10 drops essential oil per ounce of carrier oil
3% dilution:	15 drops essential oil per ounce of carrier oil

The size of a drop varies depending on the size of the dropper opening, as well as on the temperature and the viscosity (thickness) of the essential oil. Generally, an ounce is regarded as 500 drops. A drugstore dropper will probably be accurate enough for your purposes.

Some people find it easier to use drops; others prefer measuring their essential oils by the teaspoon. Teaspoons are usually more convenient when preparing large quantities. Whatever your preference, use the chart on the following page as a general guideline.

Measurement Conversion Chart

10 drops	1/10 tsp.	1/96 oz.	1/8 dram	about 1 ml.
12.5 drops	1/8 tsp.	1/48 oz.	1/6 dram	about 5/8 ml.
25 drops	1/4 tsp.	1/24 oz.	1/3 dram	about 1 1/4 ml.
50 drops	1/2 tsp.	1/12 oz.	2/3 dram	about 2 1/2 ml.
100 drops	1 tsp.	1/6 oz.	1 1/3 drams	about 5 ml.
150 drops	1 1/2 tsp.	1/4 oz.	2 drams	about 13.5 ml
300 drops	3 tsp.	1/2 oz.	4 drams	about 15 ml.
600 drops	6 tsp.	1 oz.	8 drams	about 30 ml.
24 teaspoons	(8 tablespoons)	4 oz.	1/2 cup	
48 teaspoons	(16 tablespoons)	8 oz.	1 cup	1/2 pint
96 teaspoons	(32 tablespoons)	16 oz.	2 cups	1 pint

Storage and Shelf Life

Store essential oils away from heat and light to preserve their freshness and potency. When stored properly, they have a shelf life of several years. The citrus oils have the shortest shelf life of all essential oils and are best used within one year. The longest-lasting oils tend to be the thick resins such as frankincense and myrrh, woods such as sandalwood, roots like vetiver, as well as other oils, including spikenard and patchouli. These oils can actually improve as they age.

Carrier oils should also be stored away from heat, light, and oxygen to prevent oxidation and keep the oil from becoming rancid. Refrigerating blends and carrier oils can greatly improve their shelf life. Adding jojoba or vitamin E to your essential oil blends can also help preserve freshness.

Keep it simple

You can create safe and effective remedies with just one or two oils. When combining essential oils in a therapeutic blend, it is best for beginners to keep it simple, using no more than one or two oils at a time. Using more than three may lead to unpredictable results because of the complex chemistry created by the combination of all the oils.

It is better to learn as much as you can about a small number of oils and know them very well than to know just a little about many oils. Because many plants have multiple actions, many oils are effective for a wide variety of treatments and conditions.

Remember that less is more when it comes to aromatherapy. Consistent low doses are much more effective than higher amounts.

One advantage of aromatherapy treatments is that they don't need to be applied directly to the affected areas to be effective. Most essential oils can be diluted and massaged physically onto the body but application through inhalation is also appropriate for many health conditions.

Herbal Preparations

When the essential oil of a plant is too strong for a person, the herb itself in a teas or tincture is likely a safe substitute. When used together whole plants and essential oils create a better potential for healing than either used alone.

Infused Oils

Oils made by macerating or steeping herbs in vegetable oils are called infused oils. To make infusions finely chop or grind fresh dried herbs and place in enough oil to cover the mixture. Let the mixture steep for a week but shake daily and by this time the oil should have taken the color, aroma, and healing characteristics of the herb. Strain the herbs out and store your infused oil in the refrigerator until used.

There are many variations on this basic recipe, use thicker oils for salves and ointments or use a lighter oil like hazelnut to make cosmetics. Don't limit yourself to medicinal uses alone, try experimenting with culinary infusions also.

Here are some basic herbal recipes:

Herbal Salves

Ingredients: 1 cup herb infused oil, ¾ ounce shaved beeswax.

Warm the oil in a pan slightly and add the beeswax until melted. Mix well and allow to cool.

Herbal Liniments

A liniment is a topic remedy, either oil based or liquid, that helps relieve pain, stiffness and sore muscles. Liniments have a long history of use and they are still very effective for muscle pain and soreness. Liniments are incredibly simple to make.

Some liniments are oil based, but my preferred liniment has an alcohol and witch hazel base with a concentration of herbs.

Alcohol or witch hazel based liniments are highly effective because the alcohol is quickly absorbed by the skin, carrying with it the beneficial properties of the added herbs.

Liniments can be made "warm" or "cool" by using different herbs. Using a mixture of cooling herbs (peppermint and menthol) and warming herbs (cayenne and ginger) can help create an alternating effect that I find effective. I also like to add herbs that help speed recovery in other ways, like arnica, comfrey and yarrow.

Ingredients

- 1 cup of witch hazel (rubbing alcohol will also work)
- 1 teaspoon menthol crystals or 2 tablespoons peppermint essential oil
- 1/2 teaspoon dried cayenne pepper
- Optional: 1 tablespoon fresh ginger, finely chopped
- Optional: 2 tablespoons dried herbs

Instructions

1. Place herbs and oils, menthol crystals and fresh ginger (if using) in a pint size glass mason jar.

2. Add the witch hazel or rubbing alcohol and make sure that all the herbs are covered.
3. Let the mixture sit for at least 4 weeks (8 is preferable).
4. Strain out the herbs and store the finished liniment in a spray bottle for easy use. If you don't have a spray bottle, store in any glass bottle or jar and use a cotton ball or gauze pad for easy application. Use as needed for sore muscles.

Liniments should only be used externally and on unbroken skin. This mixture is shelf stable and will last several years. If you have sensitive skin or are using this on children, reduce the amount of herbs and menthol by half. Menthol crystals will easily dissolve in alcohol or witch hazel, so if you aren't sure how strong you want your liniment, start with less and add more to the finished liniment if needed. Check with a doctor before using if pregnant/nursing or if you have a medical condition.

The reason alcohol is often used is that it penetrates skin and evaporates very quickly. It is also very effective at extracting the beneficial properties of the herbs used in this liniment and is used as an antiseptic (a liniment made with alcohol could be used on open wounds depending on the herbs used). Some people prefer not to use alcohol or find it drying, and witch hazel offers another great alternative with beneficial properties of its own.

Herbal Tea, Infusions, or Decoctions

To make an herbal tea infusion simply pour boiling water over fresh or dried herbs. Cover them and let them steep about 10 minutes, then strain. Covering helps preserve the volatile essential oils.

Solid perfumes or Unguents

Ingredients: beeswax, jojoba oil, essential oils

For this recipe use 30ml of essential oil for every 100 ml of base. To make a hard product use 4 parts jojoba to 1 part beeswax and to make a softer product use 5 parts jojoba to one part beeswax. Melt the beeswax in a saucepan and add the warm jojoba oil. Pour the warm mixture into a clean jar, add the essential oils and quickly stir before the product begins to solidify. Cover and set for 20 minutes to cool.

Lip Balm

Ingredients: beeswax, essential oils, jojoba oil or infused oils optional

Melt one ounce of beeswax in jar placed in a sauce pan about 1/3 full of water, and add a spoonful of infused oil or jojoba oil if so desired. Pour the warm mixture into a clean container and add 10 ml of essential oils. Stir quickly until product begins to cool. Cover and let cool for 20 minutes.

Clay Packs, masks, and poultices

Clay is one of the oldest of all skin treatments and is a wonderful medium for aromatherapy treatments. Clay is a balancer and revitalizer and when applied to the skin circulation is stimulated and defense functions improved. Healing clays can be purchased at health food stores, do not use clay for ceramics or crafts because those clays contain bacteria which may be harmful to the skin.

To make a clay facial mask measure out some clay into a ceramic or metal bowl and add enough warm water to form a soft paste. A few drops of essential oil diluted in a base oil complete your clay mixture. Spread the mask out across the skin area to be treated and leave the mask on until the clay is dry, then scrub it off or wash it off with warm water.

This same recipe can be used as a drawing poultice for removing splinters and drawing out infections. Simply place the clay paste directly over the area and leave the clay on overnight.

Dry clay can be mixed with baking powder (1 cup of clay to one tablespoon of baking powder) and a few drops of essential oils to make a deodorizing foot powder.

Dry clay, cornstarch, or arrowroot can all be used as the base for baby powders or foot and body powders. Common mixes call for ¼ cup cornstarch or arrowroot with 1 tablespoon of dry clay and 3 drops of essential oil.

Aromatic Baths

Essential oils should not be added undiluted to the bath. Essential oils are hydrophobic, which means they don't mix well with water, and in the bath where the only oil medium is your skin and warm water is increasing absorption rates, undiluted essential oils can penetrate much quicker than diluted oils. Mix your bath oils at a 4% dilution, or about 20-25 drops per ounce of carrier oil. Add 15 drops of this blend per tub, depending upon the essential oils being used.

Aromatic bath vinegars can be made by using 25 drops essential oil to 4 ounces of vinegar. Shake this mixture every day for a week then add 2 tablespoons to your bath.

Bath Salts

Aromatic bath salts are wonderful to use and simple to make. Bath salts are made from very simple ingredients, such as Epsom salts, sea salts, baking soda, or even table salt. A simple recipe that makes a luxurious bath is to mix ½ cup of Epsom salt, ½ cup of sea salt, and 50 drops of essential oil. Mix this thoroughly and add ¼ cup to a warm bath.

Facial Cleansers

The first facial cleanser was invented by Galen in 150 AD. It was made of olive oil, rosewater and melted beeswax. It was called cold cream because it had a cooling effect on the skin as it cleaned. A basic facial cleanser can be made of olive oil and essential oil in a 1% dilution.

Additional recipes call for 1 tablespoon of olive oil, 1 tablespoon of glycerin, and ¼ cup of water or aloe gel. Aloe gel will make a much thicker cleanser than water. Apply with fingers or cotton balls and then rinse.

Facial Toners

A simple and effective toner can be made by mixing ½ cup witch hazel with 2 tablespoons of water and 5 drops of essential oil. ¼ cup of dried herbs can be steeped in the witch hazel for 10 days to make a toner using herbs. Strain well and shake before each use. Hydrosol are also extremely effective as toners.

Creams and Lotions

It is actually quite simple to make your own creams and lotions in your kitchen. The best preservatives for your products are the essential oils, so under most conditions your creams and lotions should be good for 6 months from the day that they are made.

The basis for the simplest and oldest forms of lotion is water, oil, and beeswax. These ingredients can be mixed and matched to any need, just maintain the proper proportion of oil to water for your recipe. All lotions and creams become harder as they cool so it is easier to keep them in wide mouthed jars. Here are the most common ingredients:

<u>Water</u>	<u>Oil</u>
Distilled water	any vegetable oil
Spring water	herbal oil infusion
Aloe Vera juice	Cocoa Butter
Hydrosols	Coconut oil

Beeswax is a very common base for homemade cosmetics. To make a basic salve, lotion, or balm use melted beeswax in a base oil and simply vary the amount of beeswax used to harden or soften the product.

Ingredients:

1/2 cup vegetable oil

1/4 cup coconut oil

1/4 cup beeswax

Optional: 1 teaspoon Vitamin E oil

Optional: 2 tablespoons shea butter or cocoa butter

Optional: Essential oils, Vanilla Extract or other natural extracts to suit your preference

Instructions:

Combine ingredients in a pint sized or larger glass jar. You can even reuse a glass jar from pickles, olives or other foods.

1. Fill a medium saucepan with a couple inches of water and place over medium heat.
2. Put a lid on the jar loosely and place in the pan with the water.
3. As the water heats, the ingredients in the jar will start to melt. Stir occasionally to incorporate. When all ingredients are completely melted, pour into whatever jar you will use for storage. Small mason jars (8 ounce) are great for this.
4. Use as you would regular lotion. This has a longer shelf life than some homemade lotion recipes since all ingredients are already shelf stable and not water is added. Use within 6 months for best moisturizing benefits.

Note: A little goes a long way! This lotion is incredibly nourishing and is also great for diaper rash on baby, for eczema and for preventing stretch marks!

Lotion Bars

Ingredients:

- 1 cup coconut oil
- 1 cup shea butter, cocoa butter or mango butter (or a mix of all three)
- 1 cup beeswax
- Optional: 1 teaspoon Vitamin E oil

Instructions:

1. Combine all ingredients (except essential oils if using) in a quart size glass mason jar with a lid and place this in a small saucepan of water until melted.
2. Turn the burner on and bring water to a boil. Stir ingredients constantly until they are melted and smooth:
3. Remove from heat and add the essential oils.
4. Gently stir by hand until essential oils are incorporated.

5. Carefully pour into molds or whatever you will be allowing the lotion bars to harden in. I use silicone baking cups, though any mold will work.
6. Allow the lotion bars to cool completely before attempting to pop out of molds. These could be made in different shaped molds for different holiday gifts (hearts for valentines, flowers for Mother's day, etc.) or made in a square baking pan and then cut into actual bars.

Note: This recipe can be adjusted to make any quantity that you'd like. I recommend starting with equal 1 cup measurements. The recipe I used made exactly 12 lotion bars with my molds. For a small batch, this recipe could be cut in half or even one fourth.

These wonderful lotion bars are solid at room temperature and look like a bar of soap. It is also even easier to make than lotion because it doesn't require any emulsifying with water, which is the tough step. These are solid at room temperature like a bar of soap, but when rubbed on the skin, a tiny amount melts and is transferred to the skin, leaving a highly moisturizing and very thin layer.

Cream

Ingredients:

- 1 cup vegetable oil
- 1 cup water
- ¾ ounce beeswax
- 30-50 drops of essential oil
- A blender, a jar, a saucepan, and a funnel

Instructions:

Melt the beeswax as before and stir in the vegetable oil until thoroughly mixed, but not too hot. You should be able to put your fingers into it without discomfort. Place the cup of warm water into a blender and place the lid on the blender, but remove the center ring. Turn the blender on and slowly pour the beeswax and oil mixture into the blender through the funnel. Consistency of the final cream will depend not only on the ingredients used, but the temperature at which they are mixed, and how steadily the oil is poured into the water.

The cream should begin to harden after about ¾ of the oil has been added. A spatula is a handy tool if you need to scrape the edges at this point, be sure to shut the blender off first. Add oil slowly until the mixture becomes too stiff to accept any more oil. You should now have a thick beautiful cream.

Add the essential oils last. You can use the blender on low to help you mix the oil in but a spatula usually works better. Scoop your cream into wide mouthed jars, and store in a cool dark place to prolong shelf life.

Natural Skin Care

What you do to your skin before you use makeup is just as important as the makeup you use and there are some great natural options for skin care.

Using natural oils to balance the skin leaves the skin soft and smooth. Hydrosols are rejuvenating toners, make-up can be made from cocoa and clay. You can also use **a natural sugar scrub** (equal parts sugar and natural oil) or **natural microdermabrasion** (baking soda) to make skin look younger. There are infinite options available to you!

Natural Foundation

Start with a base of arrowroot powder and zinc oxide (you can also use cornstarch, but arrowroot works better) then slowly add in cocoa powder and finely ground cinnamon powder until you get a shade close to your skin tone. You can then store in a jar or old powder container and use a brush to apply. It may take a few tries of mixing to get the color correct for your skin tone, but most days, a quick brush of this is all you need. Adding gold mica powder gives it an even smoother texture and makes skin look radiant.

Ingredients:

- 2 tablespoons zinc oxide (Can use arrowroot powder instead if desired, but it will not offer quite as much coverage)
- 1 tablespoon arrowroot powder (optional)
- 1 teaspoon gold mica dust
- 1/2 – 1 tsp of desired natural clay powder (optional) I use a pinch of White cosmetic

clay, Fuller's Earth Clay or French green clay
- up to 1 teaspoon finely ground cocoa powder OR bonze mica powder (or both) to get desired color
- Optional: 1 teaspoon of translucent mica powder can help for really oily skin

Instructions:

1. Mix all ingredients to get desired color and coverage.
2. Zinc oxide will give coverage and matte finish.
3. Colored mica powders, natural clays and cocoa powder will give color. Start slowly and add as needed, testing on inner-arm as you go to find your shade.
4. Store in a small glass jar with a lid.

Note: If you prefer, just arrowroot powder (or white cosmetic clay), cocoa powder, and (optional) cinnamon powder can be used. This will create a great and completely natural/edible foundation but it will not stay as long as a powder containing zinc oxide and mica.

There is a lot of variation in this recipe… creating homemade makeup is more of an art than a science and your exact color and base combinations will depend on the amount of coverage and color you want. Zinc oxide as a base will create a makeup very similar to store bought mineral makeups and adding mica powder will give skin a natural "glow."

Scrubs and Exfoliants

Exfoliating skin treatments have been used for centuries by nearly all cultures around the globe to slough off dead, dry skin cells and leave the skin looking healthy and new. The most common type of scrubs used in expensive spas and salons around the world can easily be made at home with a few simple ingredients.

Salt Scrubs

Salt scrubs can be made using any type of salt that use choose. Larger salts like sea salt crystals can provide a very aggressive exfoliation and should only be used on thicker, darker skin. Smaller salt crystals, like those found in table salts, will provide less exfoliation and are more applicable to sensitive skin and light skin complexions.

Ingredients:

- 1 cup of salt
- 1 tablespoon essential oil
- 1/2 cup vegetable oil
- Optional: 2 tablespoons dried herbs

Instructions:

Mix the ingredients together and store in a clean container for up to 6 months. To use this scrub, moisten the area to be treated (hands or feet) and then take a spoonful of the product and work it into the skin in small circles. After massaging the desired amount rinse the product off with warm water.

Sugar Scrubs

Many types of skin are too sensitive to tolerate salt scrubs as a daily exfoliant. Additionally, salt cannot be used on skin that has been freshly shaved, waxed, or is sunburned or irritated. Salt also stings if you get it into a wound. For this reason, sugar scrubs have become very popular. Sugar scrubs provide a less intensive exfoliation that can be used daily on any skin type without stinging or burning. The instructions are identical to those for making salt scrubs, just substitute a form of sugar for the salt.

Hydrosols

When essential oils are created using steam distillation, the water that is used as steam becomes impregnated with the hydrophilic compounds of the plant that are not present in essential oils. Hydrosols have incredible anti-inflammatory properties and are very soothing to the skin, and are also slightly astringent, which makes them perfect toners and cleansers for the skin. They are wonderful toners by themselves or they can be added to other ingredients like Aloe Vera. Hydrosols are valuable ingredients in masks, lotions, and creams and can be used in place of water in most aromatherapy applications to boost the health benefits.

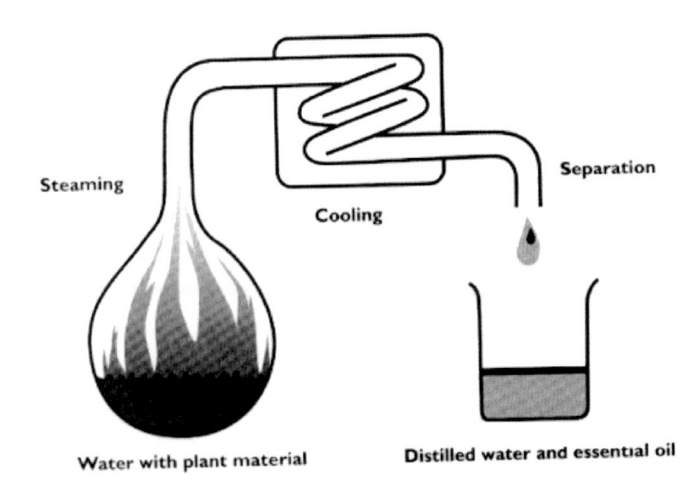

Steaming — **Cooling** — **Separation**

Water with plant material — Distilled water and essential oil

Fragrant Waters

Fragrant waters are made by adding essential oils directly to purified water. They are inexpensive and extremely easy to make and use, but do not contain the same hydrophilic compounds that are found in true hydrosols.

Fragrant waters are so simple to make and have so many uses this is one of the most popular ways for people to use essential oils in their home. Fragrant waters can be used as disinfectants, cleaners, deodorizers, antibacterial sanitizer, or as an instant aromatherapy pick-me-up.

Coloring and Texture

Enhancing your aromatherapy treatments with colors and textures is easy to do and often improves the therapeutic effects of your products by incorporating whole plant material in addition to essential oils.

Dried herbs and flowers, dried seeds, colored clays, cocoa powder, cinnamon, and even food coloring can be added to any of your creations to add color, life, and as therapeutic ingredients.

Essential Oil Recipes

There are an incredible number of essential oils available on the market that are applicable for aromatherapy products and treatments. Each of these oils is incredibly complex, and has many different uses and effects. Some oils produce physiological effects, some oils produce psychological effects, and some produce both physiological and psychological effects.

When oils are combined you can receive the effects of both of those oils, but the combination of chemical components can actually produce different effects than either of the oils would alone. As the number of oils increases the complexity of the chemical reactions created and the effects experienced also rapidly changes, and these changes become unpredictable if too many different types of oils are used.

Learn as much as you can about a few select essential oils, and by using them in different ways and in different combinations, a wide variety of ailments and conditions can be treated, a great number of wonderful products can be created, and you can safely benefit from the wonderful healing properties of fragrance.

For an example, and to get you started, I have included descriptions and uses for 5 of the commonly used essential oils around the world today. By no means should you feel limited by the selection of oils in this guide, do your research and determine the best oils for you and your desired goals. Once you have found oils you like try several different producers of that oil to find the one that works best for you.

Peppermint

Peppermint essential oil has so many uses, it's no wonder this oil is a staple in many people's medicine cabinets. It's used to freshen the breath, soothe nausea, IBS, and other stomach issues, increase alertness, improve memory and mood, and cool overworked muscles. Peppermint also helps to clear congestion, quiet headaches and tackle symptoms from PMS.

It is important to note that peppermint essential oil is intense and very concentrated. It is necessary to dilute with a carrier oil like almond or jojoba oil. It is never advised for use on young children under 6.

The entire plant contains menthol, an organic compound with local anesthetic properties which provides a wonderful cooling sensation while it naturally alleviates discomfort. Peppermint oil can be up to 40%

menthol. It also has antibacterial, antiviral, anti-inflammatory, insecticidal, antispasmodic and carminative properties.

The health benefits of peppermint oil include its ability to treat:

Indigestion: One of the oldest and most highly regarded herbs for soothing digestion is peppermint. Peppermint oil is very helpful in digestion and to relieve flatulence or abdominal pain.

- Massage several drops on your abdomen, place a drop on wrists, or inhale to soothe motion sickness or general nausea.
- Drinking mint tea has long been the antidote to an upset stomach.
- Inhaling peppermint oil is also said to help curb the appetite by triggering a sense of fullness.

Colds/Congestion: Menthol provides effective relief from many respiratory problems including nasal congestion, sinusitis, asthma, bronchitis and the common cold and cough. It is often included as an ingredient in natural chest rubs to help with congestion.

- Massage 2-3 drops (along with a carrier oil) onto chest or drop into a humidifier to help clear sinus and lung congestion.

- If your head is feeling stuffed or you can't stop coughing, try a peppermint essential oil steam. Pour boiling water into a metal or glass bowl, and add a few drops of essential oil (eucalyptus and rosemary are good combinations with peppermint). Drape a towel over your head and position your face 10-12 inches above the bowl and breathe in the steam.

Headache: Peppermint oil is a very effective treatment for some types of headaches or migraines. The use of this oil has also been known to effectively lessen tandem symptoms such as nausea, vomiting, sensitivity to noise and sensitivity to light.

- With a small amount of almond or other carrier oil and a drop of peppermint oil, rub on your temples, forehead, over the sinuses (avoid contact with eyes), and on the back of the neck to help soothe headache and pressure. When applied topically, peppermint oil leaves a soothing, cooling sensation that tends to work wonders.

Stress: Like many other essential oils, peppermint is able to provide relief from stress, depression and mental exhaustion. It is also effective against feeling anxious and restless.

- For stress relief, the combination of peppermint, lavender and geranium essential oils added to a warm bath help relieve stiffness while absorbed through the skin.
- Try aromatherapy by using a diffuser with essential oils.

Energy/Alertness: Peppermint oil powerfully affects and improves mental clarity and raises energy levels. If you're trying to cut back on caffeine, this may be a helpful alternative.

- Rub a drop of oil under the nose to help improve concentration and alertness.
- Diffuse Peppermint oil in the room to improve concentration and accuracy.
- Apply to the back of the neck and shoulders repeatedly to keep energy levels up during the day.
- Inhale before and during a workout to help boost your mood and reduce fatigue.

Sore Muscles: Because peppermint oil has analgesic, anti-inflammatory and anti-spasmodic properties, it can relieve pain and inflammation and also calm the spasms that cause muscle cramps.

- For bug bites, use a combination of peppermint essential oil and lavender essential oil to quickly dismiss the itch! Remember to dilute with a carrier oil if you are sensitive to straight essential oil on your skin.

- Add some peppermint oil to shampoo to treat dandruff.
- If you have a problem with ants in your house, leave a peppermint soaked cotton ball in their pathway. They're not big fans of mint and you'll have the nice aroma lingering in your home!
- For tired aching feet, add a few drops to a foot bath for some relief of sore, swollen and overworked feet!
- Give your trash can area a break and add a few drops to the bottom for a pleasant minty aroma.

Eucalyptus

The health benefits of eucalyptus oil are well-known and wide ranging, and its properties include anti-inflammatory, antispasmodic, decongestant, deodorant, antiseptic, antibacterial, stimulating, and other medicinal qualities. Eucalyptus essential oil is colorless and has a distinctive taste and odor.

Though eucalyptus essential oil has wonderful properties as a volatile oil, little was known about it until recent centuries. The numerous health benefits of eucalyptus oil have attracted the attention of the entire world, and it has stimulated a great deal of exploration into its usage in aromatherapy as well as in conventional medicine.

Eucalyptus essential oil is obtained from fresh leaves of the tall, evergreen eucalyptus tree. The tree, scientifically classified as *Eucalyptus Globulus* is also known as fever tree, blue gum tree or stringy bark tree, depending on where it is located in the world.

Eucalyptus is native to Australia and has spread in the past few centuries to other parts of the world including India, Europe and South Africa. Though many countries produce eucalyptus oil in small quantities, the prime source of eucalyptus oil for the world is still Australia. It is used in a variety of over the counter drugs including rubs, inhalers, liniments, rash creams, and mouthwashes.

The health benefits of eucalyptus oil include the following:

Wounds: Eucalyptus essential oil has antiseptic qualities and is used for healing wounds, ulcers, burns, cuts, abrasions and sores. It is also an effective salve for insect bites and stings. It also protects the open wound or irritated area from developing infections.

Respiratory problems: Eucalyptus essential oil is effective for treating a number of respiratory problems including cold, cough, runny nose, sore throat, asthma, nasal congestion, bronchitis and sinusitis. Eucalyptus oil is antibacterial, antifungal, antimicrobial, antiviral, anti-inflammatory and decongestant. Patients suffering from non-bacterial sinusitis showed faster improvement when given medicine containing eucalyptus oil. Gargles of eucalyptus oil mixed with warm water are consistently effective in treating sore throats.

Asthma is a condition that affects millions of people around the world, and there are many known treatments for the condition. One of these is the use of eucalyptus essential oil. Simply massage 1-3 drops onto the chest and the soothing effect of the aroma and vapors will calm the throat and dilate the blood vessels, which will allow more oxygen into the lungs. The anti-inflammatory properties of eucalyptus essential oil open bronchial passages and give relief from asthmatic symptoms.

Mental exhaustion: Eucalyptus oil is a stimulant, helping alleviate exhaustion and mental sluggishness and rejuvenating the sick. It can also be effective in the treatment of stress and mental disorders.

Eucalyptus essential oil is commonly used to stimulate mental activity and increase blood flow to the brain. This essential oil is a vasodilator and it increases the blood flow around the body by relaxing the blood vessels and allowing more blood to circulate. More blood to the brain means more brain power, and eucalyptus essential oil is commonly employed in classrooms as a form of causal aromatherapy to increase student performance. Further research must be done in this area, but all signs point to a positive correlation between brain function and eucalyptus essential oil.

Muscle pain: The volatile eucalyptus oil is analgesic and anti-inflammatory. It is often recommended to patients suffering from rheumatism, lumbago, sprained ligaments and tendons, stiff muscles, aches, fibrosis and even nerve pain. The diluted oil should be massaged on the affected areas of the body.

Dental care: Eucalyptus essential oil is very effective against cavities, dental plaque, gingivitis and other dental infections owing to its germicidal properties. This is why eucalyptus essential oil is so commonly found as an active ingredient in mouthwash, toothpaste, and other dental hygiene products.

Lice: Due to its well-known qualities as a bug repellent and natural pesticide, it is frequently used as a natural treatment of lice. Some of the main treatments of lice can be very severe and damaging to the hair, as well as packed with dangerous chemicals that you don't want being absorbed into your skin, so combing a few drops of eucalyptus oil through a lice-infested head is a much better and healthier solution.

Intestinal germs: Eucalyptus oil is frequently employed to remove germs in the intestine. Studies have shown that ingesting eucalyptus oil can deter many of the bacterial, microbial, and parasitic conditions that affect the body, particularly in susceptible areas like the colon and intestine.

Skin care: Eucalyptus oil is often applied topically to treat skin infections.

Diabetes: Eucalyptus oil can help control blood sugar. Therefore, it is a good idea for diabetics to frequently massage eucalyptus oil onto the skin topically, and also inhale it as a vapor.

Lavender

The health benefits of lavender essential oil include the ability to eliminate nervous tension, relieve pain, disinfect the scalp and skin, enhance circulation and treat respiratory problems. The Latin name of lavender is *Lavare*, which means "to wash", due to its aroma.

Lavender oil is extracted mostly from the flowers of the lavender plant, primarily through steam distillation. The flowers of lavender are fragrant and have been used for making potpourri, perfumes, and medicines for centuries. The oil is very useful in aromatherapy and many aromatic preparations and combinations are made using lavender oil.

Today, lavender essential oil is frequently used in various forms including aromatherapy oil, gels, infusion, lotion, and soaps.

Health Benefits of Lavender Essential Oil

The various health benefits of lavender essential oil include the following:

Bug Repellent: The smell of lavender essential oil is potent for many types of bugs like mosquitoes, midges, and moths. Apply some of the lavender oil on the exposed skin when outside to prevent these irritating bites. Furthermore, if you do happen to be bitten by one of those bugs, the lavender essential oil has anti-inflammatory qualities that will reduce the irritation and the pain associated with bug bites.

Sleep: Lavender essential oil induces sleep so it is a common recommendation for treatment of insomnia. Frequent studies on elderly patients have shown an increase in their sleep regularity when their normal sleep medication is replaced with some lavender essential oil placed on their pillows. It can often replace modern medicine for sleep issues.

Nervous system: Lavender essential oil has a calming scent which makes it an excellent tonic for the nerves and anxiety issues. It can also be helpful in treating migraines, headaches, depression, nervous tension and emotional stress. The refreshing aroma removes nervous exhaustion and restlessness while also increasing mental activity. It has a well-researched impact on the autonomic nervous system, which is why it is frequently used as a treatment for insomnia and also as a way to regulate heart-rate variability.

Acne: According to dermatologists and aromatherapists, lavender essential oil is one of the most beneficial oils in the treatment of acne, which is a condition that primarily affects young people as they move through puberty, but can also afflict adults. It is characterized by red, raised sores on the face and body that develop due to a bacterial infection near the sebum gland. When sebum cannot be properly secreted from the sebum glands on the face it begins to build up, and bacteria feeds off of it, creating a vicious cycle of irritation, infection, and visible sores that can result in serious scarring.

    Lavender essential oil inhibits the bacteria that cause the initial infection, helps to regulate some of the over-excretion of sebum by hormonal manipulation, and can reduce the signs of scarring after the acne has begun to heal. Adding a small amount of lavender essential oil to other skin creams or ointments can greatly increase the potential for relief and healing.

Pain relief: Lavender essential oil is known as an excellent remedy for various types of pains including those caused by sore and tense muscles, muscular aches, rheumatism, sprains, backache and lumbago. A regular massage with lavender oil can also provide relief from pain in the joints. A study done on postoperative pain relief showed that combining lavender essential oil vapor into the oxygen significantly reduced the amount of pain experienced, versus those patients only revived with oxygen after a major surgery.

Urine flow: Lavender essential oil is good for urinary disorders because of its stimulating effect on urine production. Furthermore, it helps in restoring hormonal balance and reducing cystitis or inflammation of the urinary bladder. It also reduces any associated cramps with these and other disorders.

Respiratory disorders: Lavender oil is widely used for various respiratory problems including throat infections, flu, cough, cold, asthma, sinus congestion, bronchitis, whooping cough, laryngitis, and tonsillitis. The oil is either used in the form of vapor or is applied on the skin of the neck, chest and back. It is also added to many vaporizers and inhalers and commonly used for colds and coughs.

Lavender essential oil can also loosen up the phlegm and relieve the congestion associated with respiratory conditions, speeding up the recovery process and helping the body naturally eliminate phlegm and other unwanted material. The vapor of lavender essential oil also has antibacterial qualities which can battle respiratory tract infections.

Hair care: Lavender essential oil is useful for hair care because it has been shown to be very effective on lice, lice eggs, and nits. Furthermore, lavender essential oil has also been shown to be very helpful in the treatment of hair loss, particularly for patients who suffer from alopecia, an autoimmune disease where the body rejects its own hair follicles. A Scottish study found that more than 40% of alopecia patients in the study reported an increase in hair growth when they regularly rubbed lavender essential oil into their scalp. Therefore, lavender oil is sometimes recommended as a preventative measure for male pattern baldness.

Blood circulation: Lavender essential oil is good for improving the circulation of blood in the body. Research suggests that aromatherapy using lavender oil has beneficial effects on coronary circulation. It also lowers blood pressure and is used for hypertension. This means that organs increase their levels of oxygenation, muscle strength and health is promoted, and brain activity has a noticeable boost. Additionally, the skin remains bright and the body is protected from the risks of heart attack and arthrosclerosis that is often associated with poor circulation.

Digestion: Lavender oil improves digestion and increases the mobility of food within the intestine. The oil stimulates the production of gastric juices and bile, thus aiding in the treatment of indigestion, stomach pain, colic, flatulence, vomiting and diarrhea.

Immunity: Regular use of lavender essential oil provides resistance to a variety of diseases. It is well-known that lavender has antibacterial and antiviral qualities that make it perfect for defending the body against rare diseases like TB, typhoid, and diphtheria, according to early research in the 20th century.

General Skin care: The health benefits of lavender oil for the skin can be attributed to its antiseptic and antifungal properties. It is used to treat various skin disorders such as acne, wrinkles, psoriasis, and other inflammatory conditions. It is commonly used to speed up the healing process of wounds, cuts, burns, and sunburns because it improves the formation of scar tissues. Lavender oil is also added to chamomile to treat eczema.

The oil can also be used to repel mosquitoes and moths, which is why you will find many mosquito repellents that contain lavender oil as one of the primary ingredients.

Cautions: Breastfeeding women should avoid using lavender essential oil. It is also recommended that patients with diabetes stay away from lavender oil. Some people may experience nausea, vomiting and headaches due to excessive use of lavender oil. Lavender oil should never be ingested, only topically applied or inhaled through means of aromatherapy or similar activities. Ingestion can cause serious health complications, characterized by blurred vision, difficult breathing, burning eyes, vomiting, and diarrhea.

Frankincense

Frankincense Oil is extracted from the gum or resin from Frankincense or Olibanum trees, whose scientific name is *Boswellia Carteri*. Frankincense has been a popular ingredient in cosmetics and incense burners for centuries. It is closely associated with religious traditions and rites.

Health Benefits of Frankincense Essential Oil

Immune System: Frankincense Oil is effective as an antiseptic, and even the fumes or smoke obtained from burning it have antiseptic and disinfectant qualities that eliminate the germs in the space where the smoke filters out. It can be applied on wounds without any known side effects to protect them from tetanus and becoming septic. It is equally good on internal wounds and protects them from developing infections.

Oral Health: Those same antiseptic qualities also make frankincense oil a useful preventative measure against oral issues, like bad breath, toothaches, cavities, mouth sores, and other infections. Look for natural oral care products that include frankincense oil if you enjoy the flavor or aroma, and want to include a strong antiseptic in your health regimen. You can even create your own all-natural toothpaste with frankincense oil and baking soda, or a mouthwash with water and peppermint oil.

Astringent: The astringent property of Frankincense Oil has many benefits, because it strengthens gums, hair roots, tones and lifts skin, contracts muscles, intestines and blood vessels, and thereby gives protection from premature losses of teeth and hair. This astringent quality also reduces the appearance of wrinkles, and combats the loss of firmness of intestines, abdominal muscles, and limbs associated with age. frankincense acts as a coagulant, helping to stop bleeding from wounds and cuts. This astringent property also helps to relieve diarrhea of various types.

Emenagogue: Frankincense essential oil reduces obstructed and delayed menstruation and delays the advent of menopause. It also helps curing other symptoms associated with menses and Post Menstrual Syndrome, such as pain in the abdominal region, nausea, headache, fatigue, and mood swings.

Reducing Scars: This is an interesting property of Frankincense Oil, and since skin health and anti-aging are such hot topics these days, this essential oil has become even more important. When applied topically or even inhaled, it can make the scars and after marks of boils, acne and pox on the skin fade at a much faster rate. This also includes the fading of stretch marks and surgical scars.

Digestive: This oil facilitates digestion, unlike common antacids which only suppress the symptoms. This oil speeds up the secretion of digestive juices (gastric juices, bile and acids) in the stomach and facilitates movement of food through the intestines by stimulating peristaltic motion. Frankincense oil eliminates gas and prevents it from building up in the body. This removal of excess gas from the intestines also gives relief from associated problems like stomach aches, pain in the abdominal region and chest, abnormal amounts of sweating, uneasiness, indigestion and many other related conditions. This means an all-around improvement in digestion and elimination.

Anti-Aging: Frankincense Oil promotes regeneration of healthy cells and also keeps the existing cells and tissues healthy. When you combine this aspect of frankincense oil with its powerful astringent capabilities, you have a potent anti-aging formula. It can help you to eliminate sun spots, remove micro-wrinkles around the eyes and cheeks, and generally tone and tighten skin all over your body, while simultaneously replacing old or dying cells with new, healthy ones!

Tonic: Frankincense essential oil tones and boosts health and is therefore considered a tonic. It benefits all the systems operating in the body, including the respiratory, digestive, nervous, and excretory systems, while also increasing strength by aiding the absorption of nutrients into the body. Frankincense oil also strengthens the immune system and keeps you protected for the future.

Diuretic: Frankincense essential oil is a natural and safe alternative to chemical diuretics. It promotes urination and helps you lose that extra water weight, as well as fats, sodium, uric acid and various other toxins from the body, with the added advantage of lowering blood pressure.

Respiratory Issues: It soothes coughs and eliminates phlegm deposited in the respiratory tracts and the lungs. Frankincense essential oil also provides relief from bronchitis and congestion of nasal tract, larynx, pharynx, bronchi, and lungs. Its antidepressant and anti-inflammatory properties also help relax the breathing passages, which can reduce the dangers of asthma attacks, and its antiseptic qualities are what give it the reputation of being an immune system booster! It also eases body pain, headaches, toothaches and balances the rise in body temperature commonly associated with colds.

Stress and Anxiety: Frankincense Oil is very effective as a sedative, and it induces a feeling of mental peace, relaxation, satisfaction and spirituality. It also awakens insight, makes you more introspective and lowers anxiety, anger and stress. Frankincense essential oil promotes deep breathing and relaxation, which can open your breathing passages and reduce blood pressure, moving your mental state back to calmness.

Uterine: This oil is very good for uterine health. Since it regulates the production of the estrogen hormone, it reduces the chances of post-menopause tumor or cyst formation in the uterus, also known as uterine cancer. In terms of the pre-menopause period, it keeps a woman's uterus healthy by regulating proper menstrual cycles. It also treats or regulates gynecologic conditions or stressors that can lead to complicated dysfunctions in certain women.

First Aid: Simply apply a diluted solution of this oil on wounds, or use it blended with a skin cream, and your wounds will heal faster and be protected from infections. This oil is equally beneficial in healing internal wounds, cuts and ulcers.

Other Benefits: It relieves pain associated with rheumatism and arthritis. It helps heal boils, infected wounds, acne, circulatory problems, insomnia and various types of inflammation.

There are no known adverse side effects. That being said, frankincense essential oil should not be used during pregnancy, since it does act as an emenagogue and astringent.

Clove

The oil of Clove is antimicrobial, antifungal, antiseptic, antiviral, and has aphrodisiac and stimulating properties. This oil is used for treating a variety of health disorders including toothaches, indigestion, cough, asthma, headache, stress and blood impurities. The most common use of clove oil is in dental care. Several toothpastes, mouth wash and oral care medications contain clove oil as an important ingredient.

Clove is an evergreen tree, which produces a flower bud that has numerous medicinal properties. It is often referred to as clove bud. Clove bud has a shaft and a head and hence it has the Latin name clavus, meaning nail. Clove was extensively used in ancient Indian and Chinese civilizations and it spread to other parts of the world, including Europe, during the seventh and eight centuries. Clove is used in several Indian and Chinese dishes.

Clove is rich in minerals such as calcium, hydrochloric acid, iron, phosphorus, sodium, potassium, and vitamin A and vitamin C.

Health benefits of clove oil

Infections: Due to its antiseptic properties, clove oil is useful for wound, cuts, scabies, athlete's foot, fungal infections, bruises, prickly heat, scabies, and other types of injuries. It can also be used for treating insect bites and stings. Clove oil is very strong in nature and should always be used in diluted form, and it should not be used by people with unusually sensitive skin. Clove and clove oil are very effective home remedies for treating sty. Sty is an inflammation on the eyelash and can be a very irritating and painful condition. Clove oil has also been shown to be preventative of other eye infections due to its antibacterial qualities.

Dental care: As mentioned above, the most prominent use of clove oil is in dental care. The germicidal properties of the oil make it very effective for relieving dental pain, tooth ache, sore gums and mouth ulcers. Clove oil contains the compound eugenol, which has been used in dentistry for many years. Gargling with diluted clove oil helps in easing throat pain and irritation. The characteristic smell of clove oil also helps to eliminate bad breath. Clove is also effective against cavities, and traditionally, in India, clove oil was added to a small cotton ball and put at the end of the tooth which has a cavity every day before going to sleep. The cavity would vanish in a few days. As a result, clove oil is added to numerous dental products and medications, including mouthwash and tooth paste. Dentists also mix clove oil with zinc oxide to prepare a white, filling material as a temporary alternative to a root canal. But be careful, clove oil is very strong and can cause burns inside your mouth if used incorrectly.

Clove oil is being used as a soothing balm on infants who are teething. In extremely diluted form, it can be applied to a baby's gums, and the antiseptic and soothing qualities of the oil can ease their pain and reduce their discomfort.

Skin care: Clove oil is often recommended for skin care, especially for acne patients. The effects are best achieved when the oil is used in liquid form and spread on a clean, dry rag. Clove oil is in many products for lessening the effects of aging, like wrinkles, sagging skin, and facial rejuvenation for the eyes.

Immune system: Clove is very useful for boosting the immune system. Its antiviral properties and ability to purify blood increases resistance to a multitude of diseases, clove essential oil scavenges the body of dangerous free radicals that cause a multitude of diseases like heart disease and certain types of cancer.

Essential oils are ranked by their capacity to fight oxidation, and clove oil has the highest ranking of any other essential oil in the world. This means that when it comes to fighting free radicals in the body, clove oil is the king.

Stress: Clove oil is aphrodisiac in nature and serves as an excellent stress reliever. It has a stimulating effect on the mind and removes mental exhaustion and fatigue. When ingested it refreshes the mind and stimulates brain function. Clove oil also induces sleep and is helpful to patients suffering from insomnia. It is useful for treating neural disorders such as memory loss, depression and anxiety.

Headache: When mixed with salt and applied on the forehead clove oil gives a cooling effect and helps in getting relief from headaches. Clove oil has many flavonoids within it which are anti-inflammatory agents. When topically applied to the temples or neck, that anti-inflammatory quality will ease the inflammation or tension that so often brings about headaches. For the same reason clove oil is used as a pain reliever on other parts of the body.

Respiratory problems: Clove oil is also an expectorant, and is frequently used to clear the nasal passages. This essential oil is a useful treatment for various respiratory disorders including coughs, colds, bronchitis, asthma, sinusitis, and tuberculosis. Chewing a clove bud is traditionally recommended to soothe sore throats.

Ear-ache: A mixture of warm clove oil and sesame oil is a good remedy for earaches.

Indigestion: Clove oil has traditionally been effective for the treatment of stomach-related problems such as hiccups, indigestion, motion sickness, and flatulence. This is once again due to the potent effects of eugenol, one of the main functional parts of clove essential oil. Clove oil is helpful in reducing nausea and vomiting and is often used for pregnancy-related morning sickness and discomfort. Using it occasionally in aromatherapy or topically applying it to pillows at night for long-term inhalation can result in these positive effects.

Circulation: Clove oil increases your body's metabolism and circulation. An increase in blood circulation typically means a reduction in tension of the blood vessels, a problem commonly associated with tension headaches. Increased circulation adds to the oxygenation of organ systems, which increases metabolism and raises organ efficiency. In diabetes, increased blood flow can help prevent some of the most dangerous side effects that can lead to amputations, and even death.

Blood purification: Clove oil also helps in purifying the blood; studies have shown that the aroma extracts can actually reduce toxicity in the blood and stimulate antioxidant activity throughout the body, thereby boosting the immune system as well as purifying platelets.

Diabetes: Along with blood purification, clove oil helps control the level of blood sugar, making it very useful to patients suffering from diabetes. Studies have shown that glucose response mechanisms are more regulated when clove oil is acting on the body's systems.

Insect Repellent: Clove oil is commonly used as a component in bug repellent and insect-repelling candles because the vapor is very potent for the olfactory senses of many insects. Traditionally, a few drops of clove oil were placed on the bedsheets at night to keep bugs away.

Clove cigarettes: Usage of clove in making cigarettes is a trend all over the world which started in Indonesia. The natural elements in clove cigarettes can reduce the harmful impacts, but smoking clove cigarettes can still be carcinogenic.

Soap: Due to its powerful aroma, soothing effect and antiseptic properties, clove oil is often added when making many types of soap.

Cautions: One should be careful while using clove oil, because it is very strong even in small quantities and must be diluted before application or ingestion. Since eugenol (a main part of clove essential oil) is not very common, some people discover violent allergies when taking too much at once. Use small amounts of any essential oil if you have never used it before. Other possible risks of clove oil include some intestinal discomfort. Clove oil can cause blood sugar to drop, so diabetics should be cautious, and pregnant women and those who are nursing should not use clove oil, as it is not clear whether this strong compound passes to the infant through breast milk.

Conclusion

This book is intended to help introduce you to the incredible world of aromatherapy and open your mind to the possibilities of using essential oils to enhance your health and your life. This book is only the beginning though, only a brief touch, of the complex and vibrant history of fragrance and the multitude of uses for these amazing oils. It is like dipping a finger into a vast ocean.

Use the information in this book as a starting point, and as your knowledge and experience grow you can expand your range of oils and create an even larger range of applications for them. In this way you will soon be able to tailor your aromatherapy products to any personal need or desire.

About the Author

Scott Schriver is a licensed massage therapist who has been preparing aromatherapy products for use and administering them professionally for over two decades. He is certified in many forms of massage including neuromuscular therapy and structural integration, and is also certified as both a functional trainer and a rehabilitation specialist. Scott specializes in integrating therapeutic bodywork, nutrition, and exercises to treat chronic painful conditions, correct compensation patterns, and improve health.

Scott began using aromatherapy in his massage practice in 1995 and taught his first class on the subject in 1997. This book is the result of decades of study, research, and application, refined through years of classroom presentation.

Scott and his wife Zoe own and operate a clinic in Saint Petersburg, Florida called Scott & Zoë's Therapy and Fitness Center. Their facility offers many massage therapy and rehabilitation services, nutritional counseling and personal training sessions, group fitness classes, and educational seminars and workshops.

In addition to working with each of his clients individually, Scott also teaches seminars, workshops, and provides services as a consultant. Seminars in nutrition, advanced massage techniques, spa services, massage modalities, hydrotherapy, aromatherapy, anatomy and physiology, and foam rolling are among those available. Scott has taught as faculty in four schools for both massage therapy and skin care and has been teaching educational seminars for over 15 years.

Free seminars on aromatherapy and other topics are available throughout the year, but fill up quickly. Check out his website at www.ScottandZoe.com to purchase some of Scott's unique aromatherapy products and leave an email address for information about upcoming classes and seminars.

Made in the USA
Columbia, SC
20 November 2020